Dear Parents and Educators,

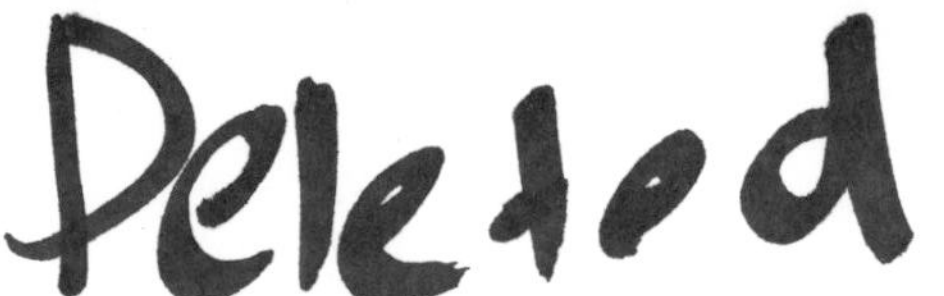

Welcome to Penguin Young Readers! As parents and educators, you know that each child develops at their own pace—in terms of speech, critical thinking, and, of course, reading. Penguin Young Readers recognizes this fact. As a result, each Penguin Young Readers book is assigned a traditional easy-to-read level (1–4) as well as an F&P Text Level (A–P). Both of these systems will help you choose the right book for your child. Please refer to the back of each book for specific leveling information. Penguin Young Readers features esteemed authors and illustrators, stories about favorite characters, fascinating nonfiction, and more!

Jellyfish!

LEVEL 4

F&P TEXT LEVEL R

This book is perfect for a **Fluent Reader** who:

- can read the text quickly with minimal effort;
- has good comprehension skills;
- can self-correct (can recognize when something doesn't sound right); and
- can read aloud smoothly and with expression.

Here are some **activities** you can do during and after reading this book:

- Comprehension: After reading the book, answer the following questions:
  - Can you name the smallest type of jellyfish?
  - What are some animals that eat jellyfish?
  - What are adult jellyfish called?
- Nonfiction: Nonfiction books deal with facts and events that are real. Talk about the elements of nonfiction. Discuss some of the facts you learned about jellyfish. Then, on a separate sheet of paper, write down facts about your favorite jellyfish from this book.

Remember, sharing the love of reading with a child is the best gift you can give!

*This book has been officially leveled by using the F&P Text Level Gradient™ leveling system.

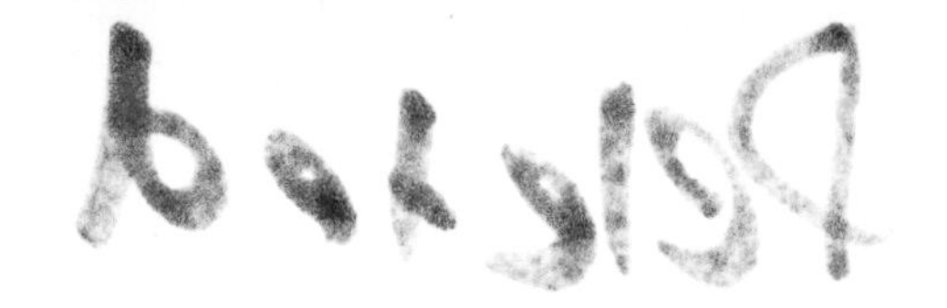

To the many students I have enjoyed meeting during author school visits, but especially for the third-graders of Tuckahoe Elementary in Richmond, Virginia, who first told me about the immortal jellyfish—GLC

PENGUIN YOUNG READERS
An Imprint of Penguin Random House LLC, New York

Photo credits: cover, 3: Freder/E+/Getty Images Plus; 4–5: Lucia Terui/Moment; 6: GondwanaGirl via Wikimedia Commons (CC BY-SA 3.0); 7: ~UserGI15667539/iStock/Getty Images Plus; 8: THOMAS PESCHAK/National Geographic Image Collection/Alamy Stock Photo; 11: Jan Bielecki, Alexander K. Zaharoff, Nicole Y. Leung, Anders Garm, Todd H. Oakley (edited by Ruthven) via Wikimedia Commons (CC BY-SA 4.0); 13: RLSPHOTO/iStock/Getty Images Plus; 14–15: Stephen Frink/The Image Bank/Getty Images Plus; 16–17: Michele Westmorland/Corbis Documentary/Getty Images Plus; 19: Damocean/iStock/Getty Images Plus; 20: Paul Sutherland/National Geographic Image Collection/Getty Images Plus; 21: olya_steckel/iStock/Getty Images Plus; 22: Searsie/iStock/Getty Images Plus; 23: SDecha/iStock/Getty Images Plus; 24: blickwinkel/Hecker/Alamy Stock Photo; 25: ttsz/iStock/Getty Images Plus; 26: GlobalP/iStock/Getty Images Plus; 27: ChrisHepburn/iStock/Getty Images Plus; 28–29: SeaTops/imageBROKER/Alamy Stock Photo; 30: NOAA Ocean Explorer via Wikimedia Commons (CC BY-SA 2.0); 32–33: LPETTET/E+/Getty Images Plus; 34–35: GaryKavanagh/iStock/Getty Images Plus; 37: Andrey Nekrasov/Alamy Stock Photo; 38: RON CHURCH/SCIENCE PHOTO LIBRARY; 39: B. Phillips (URI), D. Gruber (CUNY/Baruch College) and OET/Nautilus Live; 40: Freder/iStock/Getty Images Plus; 41: Comstock/Stockbyte/Getty Images Plus; 42–43: The Asahi Shimbun/Getty Images Plus; 43: bhofack2/iStock/Getty Images Plus; 44–45: Images & Stories/Alamy Stock Photo; 46–47: MarineMan/iStock/Getty Images Plus; 48: xijian/E+/Getty Images Plus

Published by Penguin Young Readers, an imprint of Penguin Random House LLC, New York.
Manufactured in China.

Visit us online at www.penguinrandomhouse.com.

Library of Congress Cataloging-in-Publication Data is available upon request.

ISBN 9780593093078 (pbk) 10 9 8 7 6 5 4 3 2 1
ISBN 9780593093085 (hc) 10 9 8 7 6 5 4 3 2 1

PENGUIN YOUNG READERS

LEVEL 4
FLUENT READER

by Ginjer L. Clarke

Introduction

What animals look like graceful underwater dancers but can kill like cobras? Jellyfish! There are more than 2,000 types of jellyfish. They come in many sizes, shapes, and colors.

Some look like floating plastic bags and others like silly Santa hats. Some are troublemakers, some glow in the dark, and some are deadly.

Jellyfish are found all around the world. They live in all oceans and many lakes, too. The heaviest jellyfish is the giant Nomura's jellyfish from Japan. It is six feet wide and weighs up to 500 pounds. That is almost as big as a grand piano!

Look Out Below!

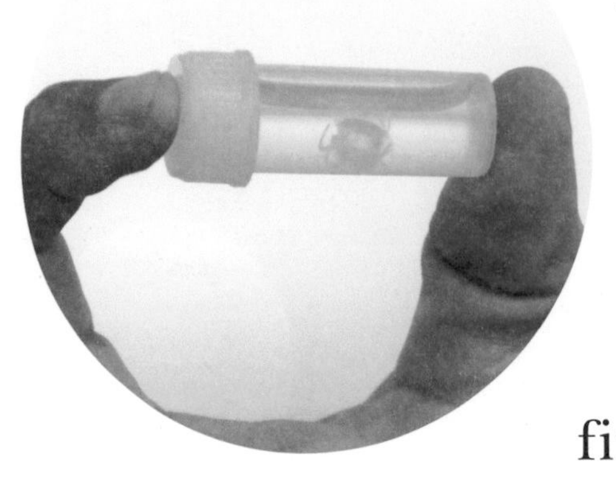

The smallest jellyfish, the kingslayer, is less than one inch wide—smaller than a fingernail. But it is deadly! Its sting is 100 times stronger than the venom of a king cobra snake!

The kingslayer is a type of box jellyfish that lives in tropical waters around Australia. Box jellies are the most dangerous of all jellyfish.

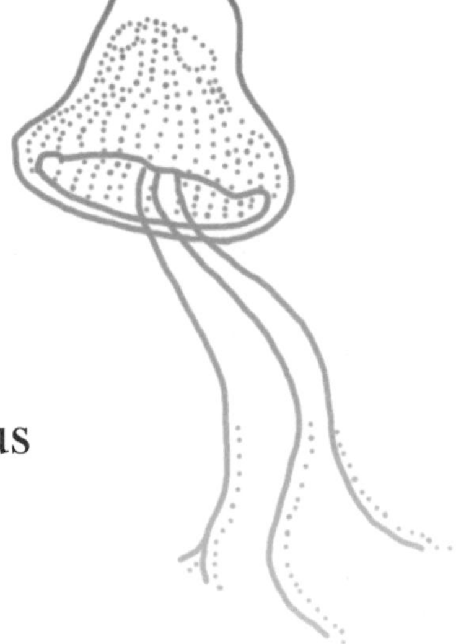

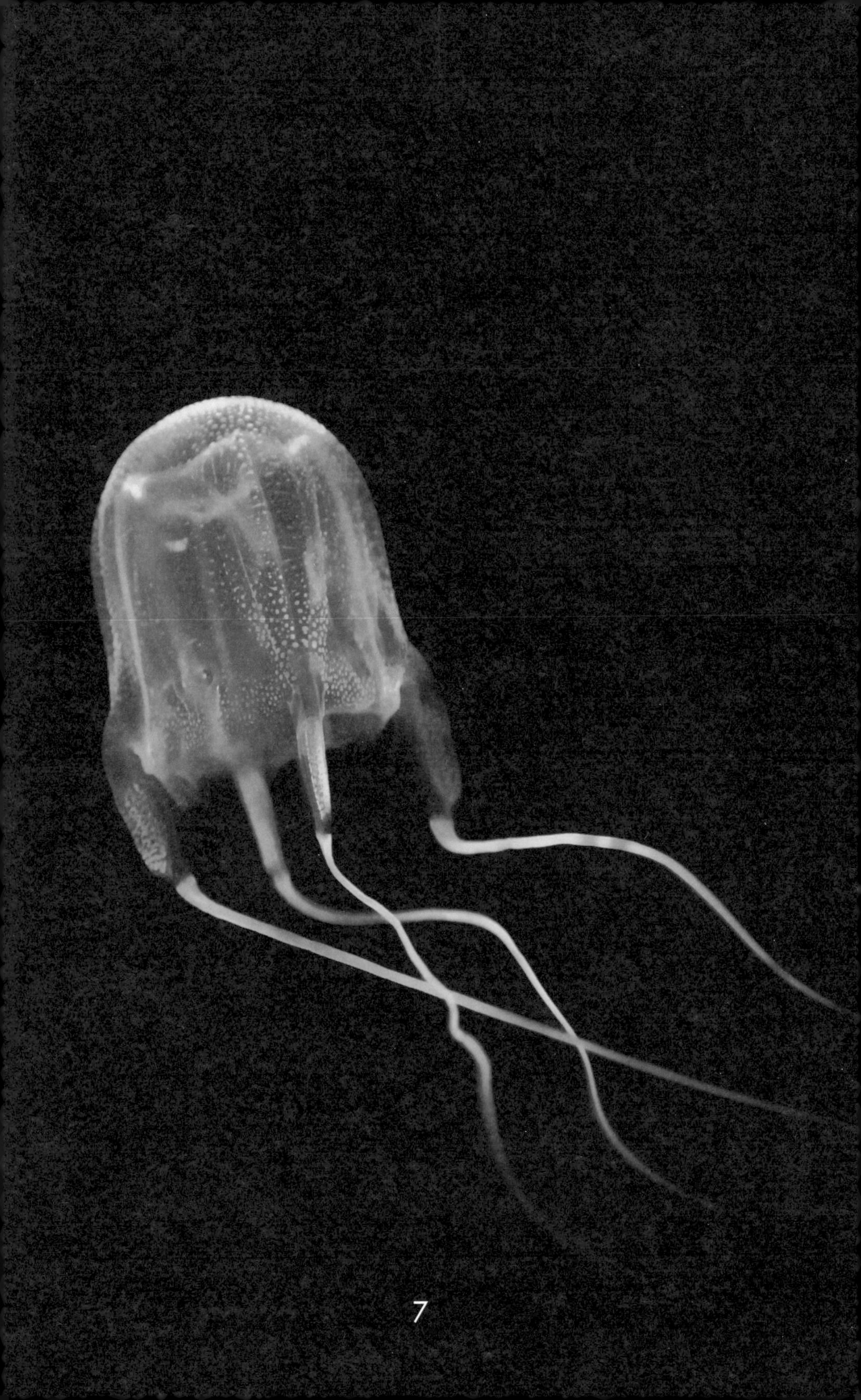

Another kind of box jelly, the Australian box jellyfish, is the most venomous animal in the world! Its sting is very painful and can kill a person in minutes.

The Australian box jelly has up to 60 long, stringy tentacles. These tentacles can sting even when they are in the water but no longer attached to the jelly's body!

Jellyfish do not attack people, though. Most stings happen when people accidentally touch a jellyfish while swimming in the ocean.

Most jellyfish cannot swim. They just float and drift. But box jellyfish can swim and hunt. *Zoom!* A box jelly sees a shrimp and moves toward it. *Boom!* The jellyfish traps the shrimp in a tentacle and stuns it with venom.

Box jellies can swim because they have something that other jellies do not—eyes. Lots of them! Box jellies have 24 eyes that they use to find their way in the water. Other jellies only have eye spots that sense light.

One jelly is so enormous that it does not need to go looking for food. It just waits for prey to come to it. This jelly's body is up to eight feet wide. Its more than 800 tentacles are each 100 feet long—longer than a basketball court! These wild, hair-like tentacles give this massive creature its name: the lion's mane jelly.

This jelly lives in the coldest waters. It spreads out its tentacles in a circle to form a net. *Trap!* It catches small fish, shrimp, and smaller jellies. *Zap!* The sting of the lion's mane jelly is so powerful that it has even been known to cause stings when it is dead!

Another animal related to the jellyfish, like a cousin, is the Portuguese man-of-war. It does not sink in the water like true jellies. Its body is like a puffy blue bag. The sail on top helps it to float like a boat.

The man-of-war is shiny and beautiful. It sparkles and shimmers in the sunlight. But its tentacles are almost invisible beneath the surface. A school of fish swims near the man-of-war's long tentacles. *Gotcha!* The fish are stung and get stuck.

Watch Us Go!

Like the man-of-war, golden jellyfish mostly stay at the water's surface. But unlike the man-of-war, golden jellies can open and close their bodies quickly to help them swim. *Whoosh! Swoosh!*

They swim slowly across Jellyfish Lake—on one of the islands of Palau

in the Pacific Ocean—as the sun moves across the sky. They need the energy from sunlight to survive.

They do not need tentacles, though, because they get energy through algae (say: AL-jee) and do not have predators. Golden jellyfish have a very mild sting. People can swim in a swarm of millions of golden jellies unharmed!

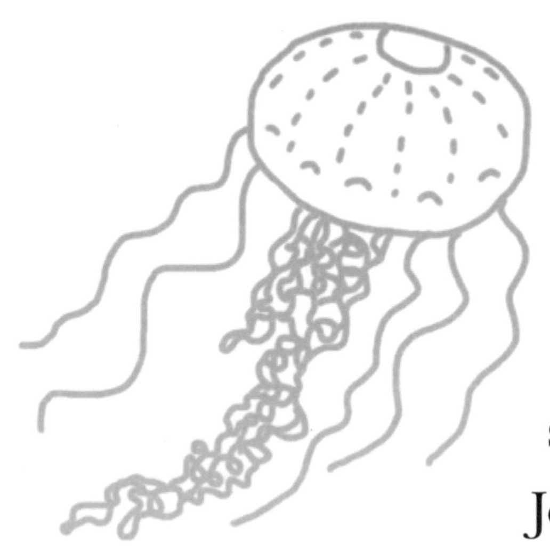

Jellyfish get their name because of their gelatinous (say: jeh-LAT-ness), or jelly-like, bell-shaped bodies.

Jellyfish cannot live without water. They breathe water like fish do. Their bodies are made up mostly of water. Unlike fish, they are invertebrates (say: in-VER-tuh-brets)—they do not have spines. They also have no brains, bones, or blood!

So how do they know where to go and what to do? They have nerve nets instead of brains. Just like people have nerves in our fingers and toes to feel things, jellies have nerve nets to sense what is around them.

Compass jellyfish

Box jellyfish

This jelly senses a fish nearby. It spreads its tentacles and waits for the fish to swim close. Then the stingers in its tentacles stun the fish. *Munch! Crunch!* The jelly grabs the fish and pops the fish into its mouth.

Some jellies eat teeny-tiny plants and creatures called plankton. Others eat larger prey, such as fish. The pink meanie jellyfish sometimes eats other jellies!

Jellyfish have very simple bodies. The mouth is the only way for food to go in and out. Jellyfish both eat and poop through their mouths. *Gross!*

Jellies are an important link in the ocean food chain. They eat many smaller creatures, but larger animals such as giant sunfish, tuna, penguins, and turtles eat them.

Sluurrp! This turtle has a mouthful of jelly. It closes its eyes to protect them from jellyfish stings.

People can eat jellies, too. Chinese fishermen have caught the edible flame jellyfish for more than a thousand years.

Eating jellyfish may become more common for people and other animals as the numbers of jellyfish continue to grow.

Most jellies begin life as an egg and then become a planula (say: PLAN-you-luh). The planula uses a sticky foot on its body to attach to a solid surface, such as a rock, a cave, or a shell. Now it is called a polyp (say: PAH-lip).

The polyp eats a lot of plankton to help it grow. Then it makes many disk-shaped copies of itself, or clones. These disks form a stack called a strobila (say: stroh-BYE-luh).

Pop! Pop! Pieces of the strobila, or ephyra (say: eh-FIE-ra), break off and swim away. They grow into adult jellies, called medusas (say: meh-DOO-suhs). Soon the adult jelly has eggs, and the cycle starts over.

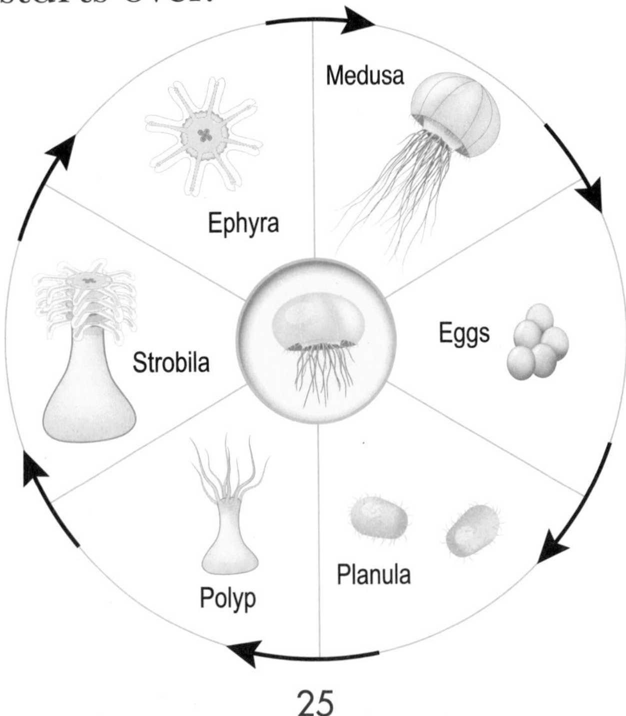

See Us Glow!

Adult moon jellies look clear and plain, but they have a cool secret. They are bioluminescent (say: BY-oh-loom-eh-NESS-ent). They glow in the dark! Fireflies and glowworms are also bioluminescent.

A moon jelly senses a big fish coming close. *Flash!* It lights up its body to confuse the fish. *Dash!* The moon jelly swims away quickly. It may also use its light show to attract another jelly and to lure prey.

Jellyfish at the surface of the water are usually colorless. But those in deep water are red or purple. These colors cannot be seen in dark water. They provide camouflage (say: KAH-muh-flazh) and help the jellies to hide. Deep-sea jellies sometimes glow, too.

The funny-looking Santa's hat jelly twinkles its red lights on and off—like Christmas tree lights. *Wink! Blink!* These lights trick a large fish into thinking the jelly is a group of smaller creatures instead of one tasty treat. So the fish leaves the jelly alone.

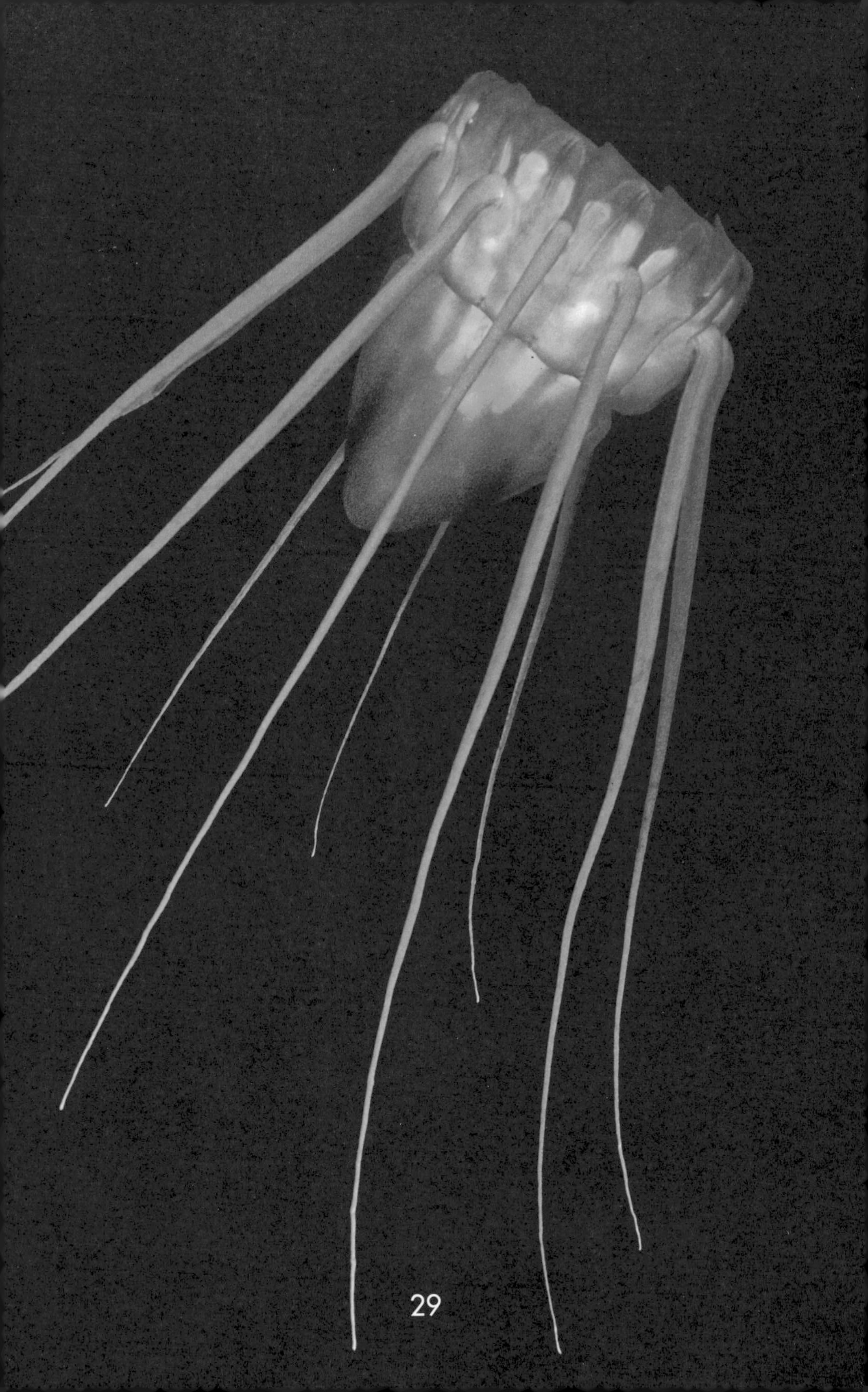

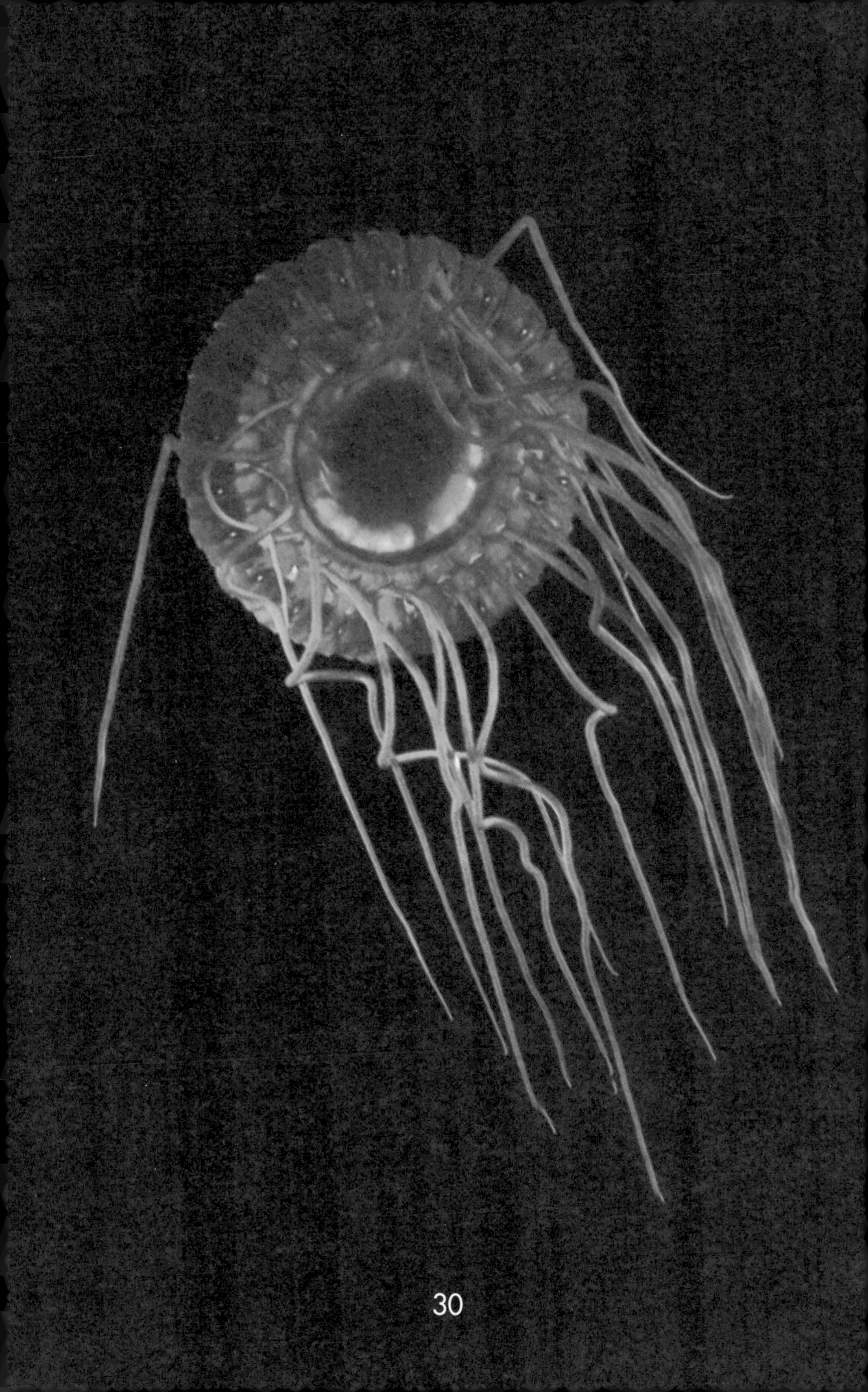

The alarm jellyfish is also dark red, and it has a fancy light display. The alarm jelly senses danger is nearby. *Whirl! Twirl!* A wave of blue lights races around its body. It is sending out an alarm signal.

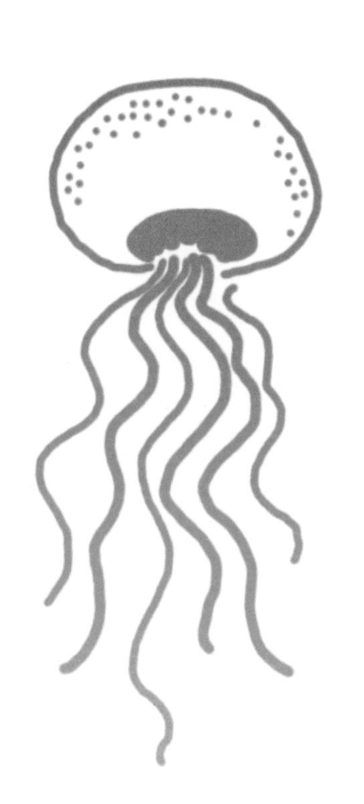

The lights are not used to scare the jelly's predator away. They are a call for help to a larger predator, such as a shark. The shark answers the jelly's alarm, and the jelly can escape.

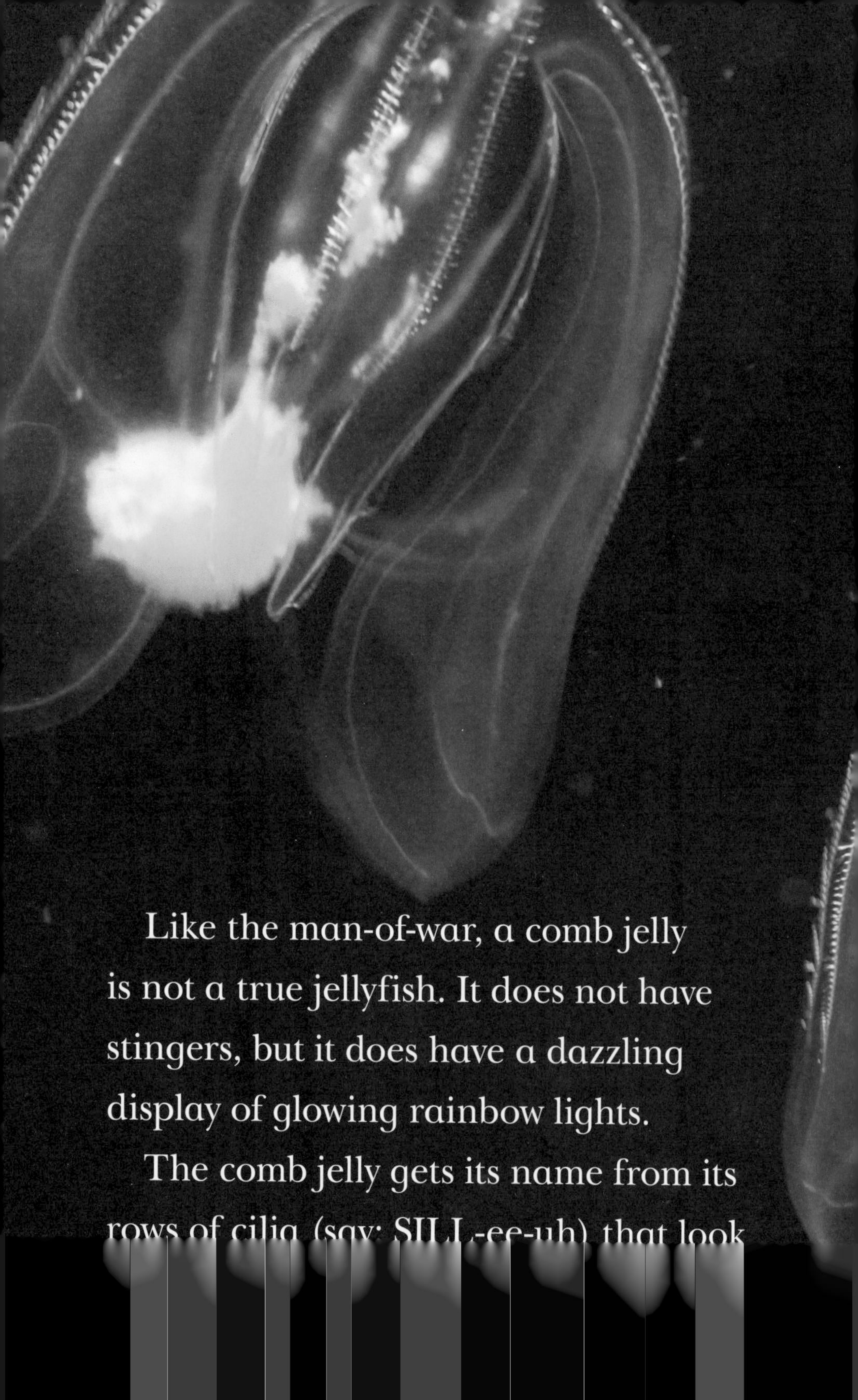

Like the man-of-war, a comb jelly is not a true jellyfish. It does not have stingers, but it does have a dazzling display of glowing rainbow lights.

The comb jelly gets its name from its rows of cilia (say: SILL-ee-uh) that look like a comb. The cilia are like hundreds of little oars that help the jelly move its body through the water.

As the cilia move, they bounce light and create rainbows that attract prey.

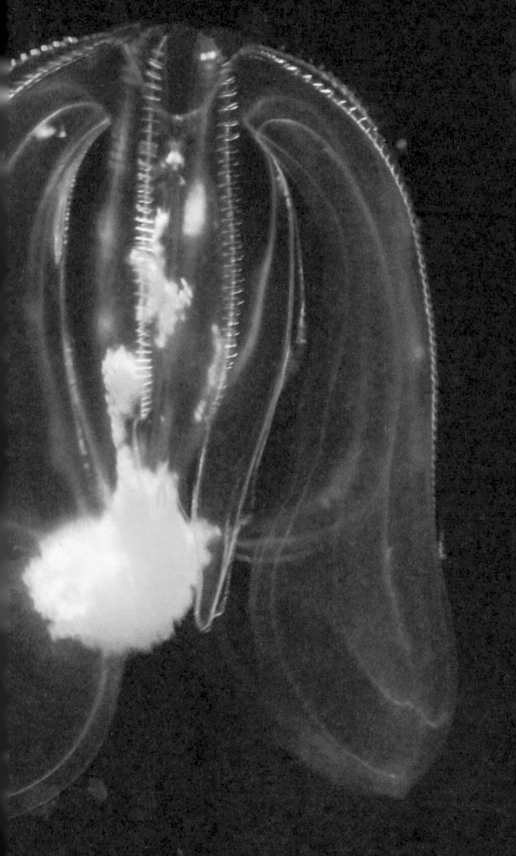

Fling! Zing! The comb jelly throws out its sticky tentacles. Its prey is caught in a flash.

The crystal jellyfish is another small jelly that can make big, bright light. It has more than 100 tiny lights surrounding its bell. These lights shine a brilliant blue-green color.

Sea creatures can see these lights, but people can only see the glow under special lighting. Some scientists discovered that the crystal jelly's glow helps us to see invisible processes such as disease that happen inside our bodies.

These scientists won an important award for what they learned using the crystal jelly's beautiful light.

What Do You Know?

The more scientists learn about jellyfish, the more they realize how little they know. For example, they once thought all jellies could swim or drift.

But a stalked jellyfish—meaning it has a stalk, or stem, like a plant—cannot move. It attaches its delicate, trumpet-shaped body to some seaweed, grass, or rocks.

Then the jelly waves its tiny, bumpy tentacles around. *Nab! Grab!* It catches a small snail, munches it slowly, and spits out the shell.

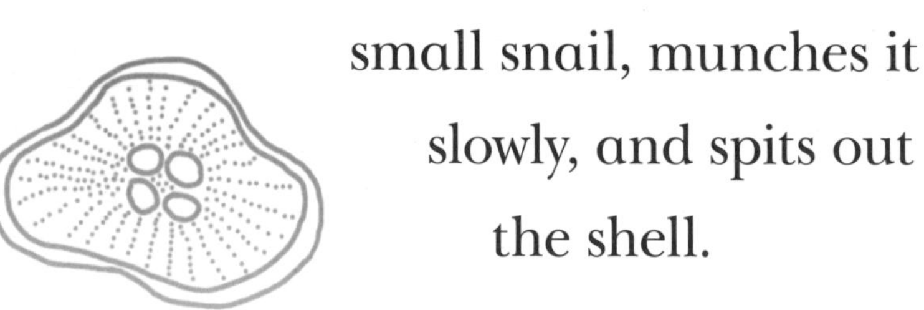

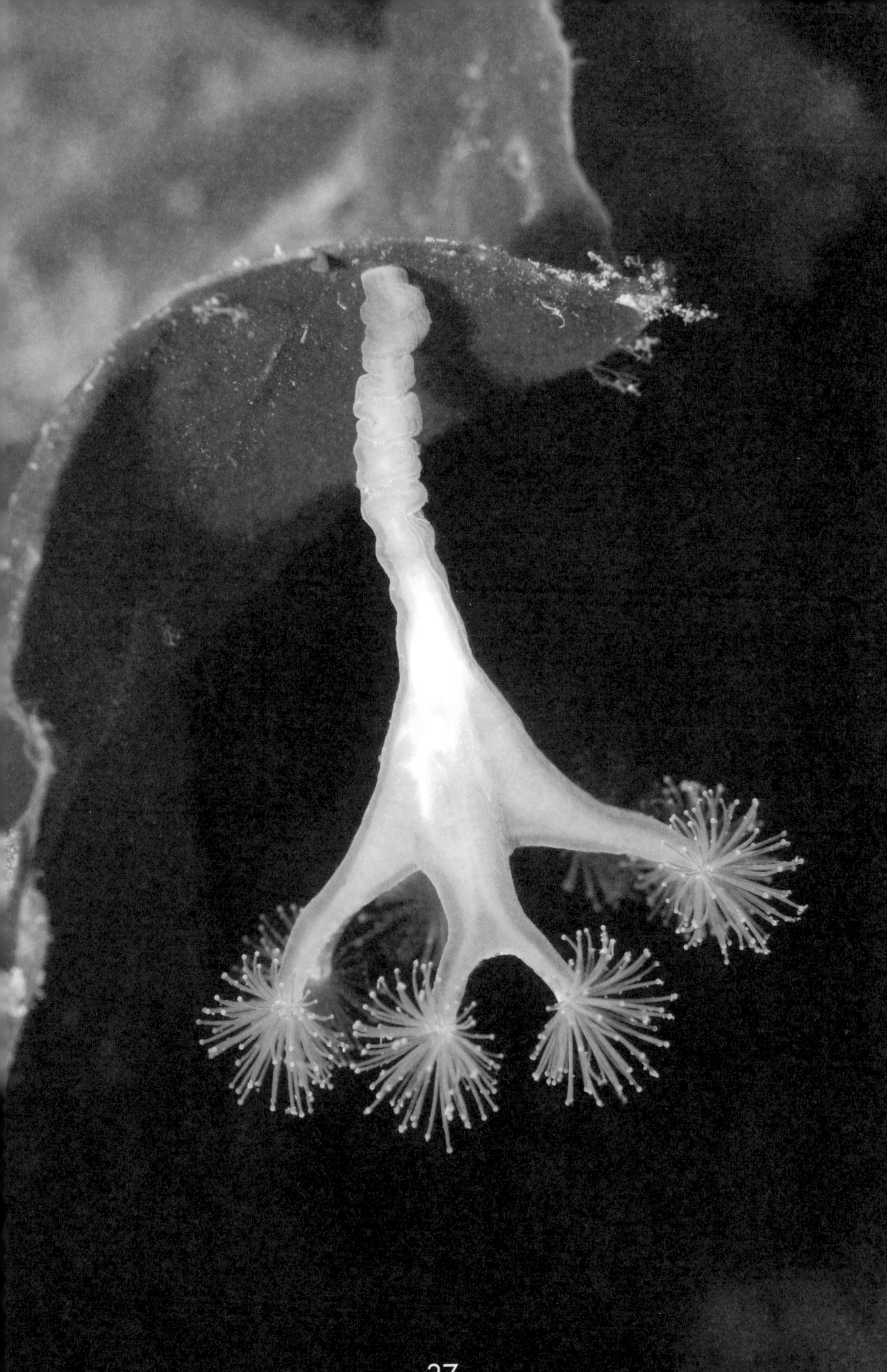

Another question scientists have wondered about is whether all jellyfish have tentacles. The jellyfish group called *Deepstaria* are named after the submarine-like boat *Deepstar 4000* that scientists were in when they first observed this type of weird jelly in the deepest part of the Pacific Ocean. They were amazed by what they saw.

This *Deepstaria* does not have tentacles, but it is still a hunter. Its huge, bag-like body flattens as it wanders in the water. *Whap!* The bumps on its body sting a crab. *Snap!* It pulls the bag closed to trap its prey.

Some jellyfish live in the deepest parts of the ocean, but others have flown high in the air. Would you believe some jellyfish have actually traveled to outer space? It's true! More than 2,000 moon jellyfish polyps were launched into space on the shuttle *Columbia* in 1991. *Blast off!*

A scientist studied the baby jellies to see how they grew without gravity and how they were different when they were brought back to Earth. This might help us understand what it would be like if humans were born in space.

USA

FK2-2315

Sometimes people use jellyfish to help solve problems, but other times the jellyfish are the problem. In the Sea of Japan, huge swarms of giant Nomura's jellyfish were tangling up fishermen's nets.

Some creative high-school students in the town of Obama, Japan, captured lots of the giant jellies in nets. And then they did something unexpected. They turned the jellies into sweet treats! The students made jellyfish powder and put it into cookies and caramel candies. *Yum!*

The most amazing discovery scientists have made about jellyfish is that one jelly can live forever!

The word *immortal* means "never dying." When the immortal jellyfish grows older, instead of dying, it starts life over. *Pow!* It pulls its tentacles back into its body, shrinks, and becomes a baby jelly again. This is like if a butterfly could turn back into a caterpillar. The only way an immortal jellyfish can die is if it is eaten. *Wow!*

This amazing ability might help scientists to better understand and one day cure diseases such as cancer.

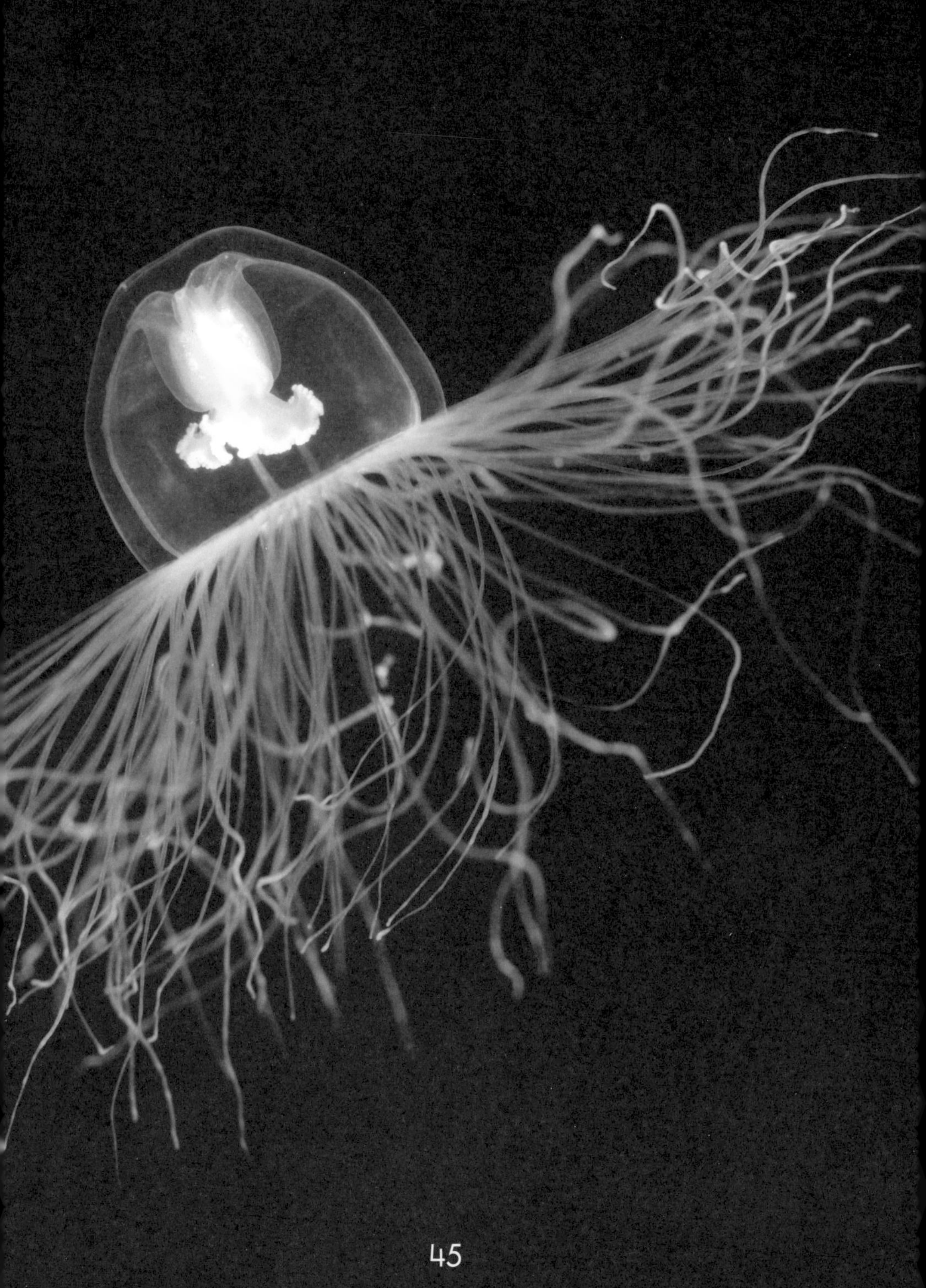

Only the immortal jelly lives forever, but jellyfish are some of the oldest creatures on Earth. They have adapted and survived for about 600 million years—and they are thriving!

Now there are more jellyfish than ever. They can live in warmer and changing oceans better than other creatures. But that can cause big trouble.

Sometimes millions of jellyfish gather in groups so thick that they clog fishing nets or keep people away from beaches. Jellies have even damaged and shut down nuclear power plants when they got sucked inside!